# *Cheap* AND *Easy!*

# WHIRLPOOL DRYER REPAIR

## 2000 EDITION

### Written ESPECIALLY for Trade Schools, Do-It-Yourselfers, and other "green" technicians!

## By Douglas Emley

**EB** **EB Publishing, Inc.**

Carson City, Nevada ● Phone / Fax toll free 888-974-1224 ● website: http://www.appliancerepair.net

The Author, the publisher and all interested parties have used all possible care to assure that the information contained in this book is as complete and as accurate as possible. However, neither the publisher nor the author nor any interested party assumes any liability for omissions, errors, or defects in the materials, instructions or diagrams contained in this publication, and therefore are not liable for any damages including (but not limited to) personal injury, property damage or legal disputes that may result from the use of this book.

All major appliances are complex electro-mechanical devices. Personal injury or property damage may occur before, during, or after any attempt to repair an appliance. This publication is intended for individuals posessing an adequate background of technical experience. The above named parties are not responsible for an individual's judgement of his or her own technical abilities and experience, and therefore are not liable for any damages including (but not limited to) personal injury, property damage or legal disputes that may result from such judgements.

The advice and opinions offered by this publication are of a subjective nature ONLY and they are NOT to be construed as legal advice. The above named parties are not responsible for the interpretation or implementation of subjective advice and opinions, and therefore are not responsible for any damages including (but not limited to) personal injury, property damage, or legal disputes that may result from such interpretation or implementation.

The use of this publication acknowledges the understanding and acceptance of the above terms and conditions.

Whirlpool®, Kitchenaid®, Roper®, Estate®, and Design 2000® are registered trademarks of Whirlpool Properties, Inc., Benton Harbor, Michigan, USA.

Kenmore® is a registered trademark of Sears, Roebuck and Co., Chicago, Illinois USA.

*Special Thanks to technical consultant Dick Miller, whose forty years of experience as a field appliance service technician are reflected in the technical information and procedures described in this publication.*

Published by EB Publishing, Inc., 251 Jeanell Drive Suite 3, Carson City, NV 89703

This service manual was excerpted from EB Publishing's Cheap and Easy! Clothes Dryer Repair Manual and edited for content.

Printed in the United States of America

# HOW TO USE THIS BOOK

**STEP 1: READ THE DISCLAIMERS ON THE PREVIOUS PAGE.** This book is intended for use by people who have a bit of mechanical experience or aptitude, and just need a little coaching when it comes to appliances. If you don't fit that category, don't use this book! We're all bloomin' lawyers these days, y'know? If you break something or hurt yourself, no one is responsible but **YOU**; not the author, the publisher, the guy or the store who sold you this book, or anyone else. Only **YOU** are responsible, and just by using this book, you're agreeing to that and more. If you don't understand the disclaimers, get a lawyer to translate them **before** you start working.

Read the safety and repair precautions in section 1-5. These should help you avoid making too many *really* bad mistakes.

**STEP 2: READ CHAPTERS 1 & 2:** Everything else in this book flows from chapters 1 and 2. If you don't read them, you won't be able to properly diagnose your dryer.

Know what kind of dryer you have and basically how it works. When you go to the appliance parts dealer, have the nameplate information at hand. Have the proper tools at hand, and know how to use them.

**STEP 3: READ THE CHAPTER ABOUT YOUR SPECIFIC MODEL.**

**STEP 4: FIX THE BLOOMIN' THING!** If you can, of course. If you're just too confused, or if the book recommends calling a technician for a complex operation, call one.

## WHAT THIS BOOK WILL DO FOR YOU
### (and what it won't!)

This book **will** tell you how to fix the most common problems with the most common brands of domestic (household) dryers. (This represents 95+ percent of all repairs that the average handyman or service tech will run into.)

This book **will not** tell you how to fix your industrial or commercial or any very large dryer. The support and control systems for such units are usually very similar in function to those of smaller units, but vastly different in design, service and repair.

We **will** show you the easiest and/or fastest method of diagnosing and repairing your dryer.

We **will not necessarily** show you the absolute cheapest way of doing something. Sometimes, when the cost of a part is just a few dollars, we advocate replacing the part rather than rebuilding it. We also sometimes advocate replacement of an inexpensive part, whether it's good or bad, as a simplified method of diagnosis or as a preventive measure.

We **will** use only the simplest of tools; tools that a well-equipped home mechanic is likely to have and to know how to use, including a VOM.

We **will not** advocate your buying several hundred dollars' worth of exotic equipment or special tools, or getting advanced technical training to make a one-time repair. It will usually cost you less to have a professional perform this type of repair. Such repairs represent only a very small percentage of all needed repairs.

We **do not** discuss electrical or mechanical theories. There are already many very well-written textbooks on these subjects and most of them are not likely to be pertinent to the job at hand; fixing your dryer!

We **do** discuss rudimentary mechanical systems and simple electrical circuits.

We expect you to be able to look at a part and remove it if the mounting bolts and/or connections are obvious. If the mounting mechanism is complicated or hidden, or there are tricks to removing or installing something, we'll tell you about it.

You are expected to know what certain electrical and mechanical devices are, what they do in general, and how they work. For example, switches, relays, heater elements, motors, solenoids, cams, pullies, idlers, belts, radial and thrust (axial) bearings, flexible motor couplings, splines, gas valves, air seals, and centrifugal blowers and axial-flow fans. If you do not know what these things do, learn them BEFORE you start working on your dryer.

You should know how to cut, strip, and splice wire with crimp-on connectors, wire nuts and electrical tape. You should know how to measure voltage and how to test for continuity with a VOM (Volt-Ohm Meter). If you have an ammeter, you should know how and where to measure the current in amps. If you don't know how to use these meters, there's a brief course on how to use them (for *our* purposes *only*) in Chapter 1. See section 1-4 before you buy either or both of these meters.

A given procedure was only included in this book if it passed the following criteria:

1) The job is something that the average couch potato can complete in one afternoon, with no prior knowledge of the machine, with tools a normal home handyman is likely to have.

2) The parts and/or special tools required to complete the job are easily found and not too expensive.

3) The problem is a common one; occuring more frequently than just one out of a hundred  machines.

More expensive repairs are included in this book only if they pass the following criteria:

1) The cost of the repair is far less than replacing the machine or calling a professional service technician, and

2) The repair is likely to yield a machine that will operate satisfactorily for several more years, or at least long enough to justify the cost.

In certain parts of the book, the author may express an opinion as to whether the current value of a particular machine warrants making the repair or "scrapping" the machine. Such opinions are to be construed as opinions ONLY and they are NOT to be construed as legal advice. The decision as to whether to take a particular machine out of service depends on a number of factors that the author cannot possibly know and has no control over; therefore, the responsibility for such a decision rests solely with the person making the decision.

I'm sure that a physicist reading this book could have a lot of fun tearing it apart because of my deliberate avoidance and misuse of technical terms. However, this manual is written to simplify the material and inform the novice, not to appease the scientist.

*NOTE: The diagnosis and repair procedures in this manual do not necessarily apply to brand-new units, newly-installed units or recently relocated units. Although they **may** posess the problems described in this manual, dryers that have recently been installed or moved are subject to special considerations not taken into account in this manual for the sake of simplicity. Such special considerations include installation parameters, installation location, the possibility of manufacturing or construction defects, damage in transit, and others.*

*This manual was designed to assist the novice technician in the repair of home (domestic) dryers that have been operating successfully for an extended period of months or years and have only recently stopped operating properly, with no major change in installation parameters or location.*

# Table Of Contents

# Chapter 1

# DRYER IDENTIFICATION
# TOOLS & SAFETY
# TIPS & TRICKS

## 1-1. BRAND IDENTIFICATION

Whirlpool has manufactured dryers for several different companies, and under several different names, including Kenmore, Kitchenaid, Estate and Roper.

## 1-2 BEFORE YOU START

Find yourself a good appliance parts dealer. You can find them in the yellow pages under the following headings:

● APPLIANCES, HOUSEHOLD, MAJOR
● APPLIANCES, PARTS AND SUPPLIES
● REFRIGERATORS, DOMESTIC
● APPLIANCES, HOUSEHOLD, REPAIR AND SERVICE

Call a few of them and ask if they are a repair service, or if they sell parts, or both. Ask them if they offer free advice with the parts they sell. (Occasionally, stores that offer both parts and service will not want to give you advice.) Often the parts counter men are ex-technicians who got tired of the pressures of in-home service. They can be your best friends. However, you don't want to badger them with TOO many questions, so know your basics before you start asking questions.

Some parts houses may offer service, too. Be careful! There may be a conflict of interest. They may try to talk you out of even trying to fix your own dryer. They'll tell you it's too complicated, then in the same breath "guide" you to their service department. Who are you gonna believe, me or them? Not all service and parts places are this way, however. If they genuinely try to help you fix it yourself, and you find that you're unable to, they may be the best place to look for service.

When you go into the store, have ready the make, model and serial number from the nameplate of the dryer.

## NAMEPLATE INFORMATION

The metal nameplate is usually found inside the door.

If it's wiped out or you can't find it, check the original papers that came with your dryer when it was new. They should contain the model number somewhere.

In any case, and especially if you have absolutely NO information about your dryer anywhere, make sure you bring your old part to the parts store with you. Sometimes they can match it up by looks or by part number.

### 1-3 TOOLS (Figure B-2)

The tools that you may need (depending on the diagnosis) are listed below.

Some are optional. The reason for the option is explained.

For certain repairs, you will need a special tool. These are inexpensively available from your appliance parts dealer. They are listed in this book as needed.

***SCREWDRIVERS:*** Both flat and phillips head; two or three sizes of each. It's best to have at least a stubby, 4- and 6-inch sizes.

***NUTDRIVERS:*** You will need at least 1/4" and 5/16" sizes. 4- or 6-inch ones should suffice, but it's better to have a stubby, too. A certain procedure when working on gas valves (Chapter 3, replacing the split coil assembly) requires a 7/32" nutdriver in most cases.

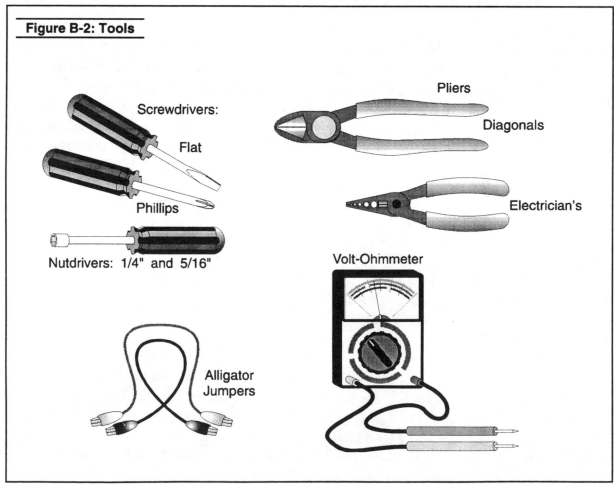

**Figure B-2: Tools**

Screwdrivers:
Flat
Phillips
Nutdrivers: 1/4" and 5/16"
Pliers
Diagonals
Electrician's
Volt-Ohmmeter
Alligator Jumpers

*ELECTRICAL PLIERS or STRIPPERS and DIAGONAL CUTTING PLIERS:* For cutting and stripping small electrical wire.

*ALLIGATOR JUMPERS (sometimes called a "CHEATER" or "CHEATER WIRE":)* Small guage (14-16 guage or so) and about 12-18 inches long, for testing electrical circuits. Available at your local electronics store. Cost: a few bucks for 4 or 5 of them.

*BUTT CONNECTORS, CRIMPERS, WIRE NUTS and ELECTRICAL TAPE:* For splicing small wire.

*VOM (VOLT-OHM METER)* For testing electrical circuits. If you do not have one, get one. An inexpensive one will suffice, as long as it has both "AC Voltage" and "Resistance" (i.e. Rx1, Rx10) settings on the dial. It will do for our purposes.

### OPTIONAL TOOLS (Figure B-3)

*SNAP-AROUND AMMETER:* For determining if electrical components are energized. Quite useful; but a bit expensive, and there are alternate methods. If you have one, use it; otherwise, don't bother getting one.

*EXTENDABLE INSPECTION MIRROR:* For seeing difficult places beneath the dryer and behind panels.

*CORDLESS POWER SCREWDRIVER OR DRILL/DRIVER WITH MAGNETIC SCREWDRIVER AND NUTDRIVER TIPS:* For pulling off panels held in place by many screws. It can save you lots of time and hassle.

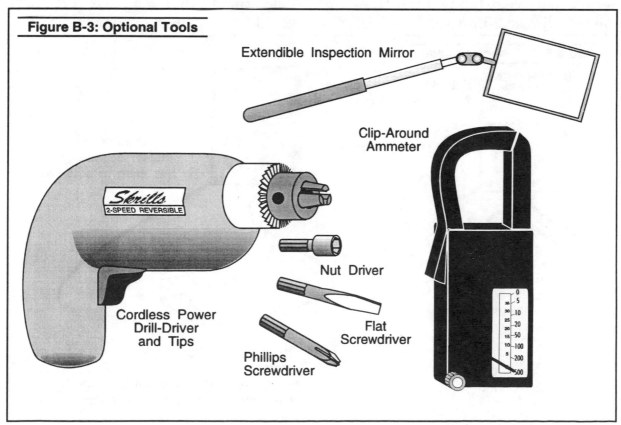

**Figure B-3: Optional Tools**

Extendible Inspection Mirror

Clip-Around Ammeter

Cordless Power Drill-Driver and Tips

Nut Driver

Flat Screwdriver

Phillips Screwdriver

## 1-4. HOW TO USE A VOM AND AMMETER

Many home handymen are very intimidated by electricity. It's true that diagnosing and repairing electrical circuits requires a *bit* more care than most operations, due to the danger of getting shocked. But there is no mystery or voodoo about the things we'll be doing. Remember the rule in section 1-5(1); while you are working on a circuit, energize the circuit only long enough to perform whatever test you're performing, then take the power back off it to perform the repair. You need not be concerned with any theory, like what an ohm is, or what a volt is. You will only need to be able to set the VOM onto the right scale, touch the test leads to the right place and read the meter.

In using the VOM (Volt-Ohm Meter) for our purposes, the two test leads are always plugged into the "+" and "-" holes on the VOM. (Some VOMs have more than two holes.)

### 1-4(a). TESTING VOLTAGE (Figure B-4)

Set the dial of the VOM on the lowest VAC scale (A.C. Voltage) over 120 volts. For example, if there's a 50 setting and a 250 setting on the VAC dial, use the 250 scale, because 250 is the lowest setting over 120 volts.

Touch the two test leads to the two metal contacts of a live power source, like a wall outlet or the terminals of the motor that you're testing for voltage. (*Do not* **jam** *the test leads into a wall outlet!*) If you are getting power through the VOM, the meter will jump up and steady on a reading. You *may* have to convert the scale in your head. For example, if you're using the 250 volt dial setting and the meter has a "25" scale, simply divide by 10; 120 volts would be "12" on the meter.

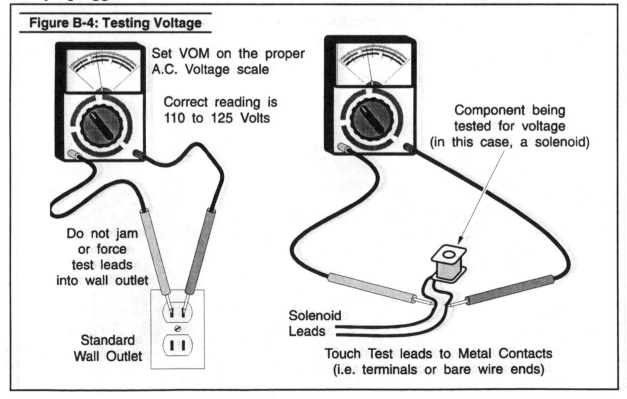

**Figure B-4: Testing Voltage**

Set VOM on the proper A.C. Voltage scale

Correct reading is 110 to 125 Volts

Do not jam or force test leads into wall outlet

Standard Wall Outlet

Component being tested for voltage (in this case, a solenoid)

Solenoid Leads

Touch Test leads to Metal Contacts (i.e. terminals or bare wire ends)

A word of caution: 220 Volts can be dangerous stuff!!! When testing 220 volt circuits (usually in electric dryers) make sure you always follow the precautions in rule 1 of section 1-5!

## 1-4(b). TESTING FOR CONTINUITY (Figure B-5)

Don't let the word "continuity" scare you. It's derived from the word "continuous." In an electrical circuit, electricity has to flow *from* a power source back *to* that power source. If there is any break in the circuit, it is not continuous, and it has no continuity. "Good" continuity means that there is no break in the circuit.

For example, if you were testing a solenoid to see if it was burned out, you would try putting a small amount of power through the solenoid. If it was burned out, there would be a break in the circuit, the electricity wouldn't flow, and your meter would show no continuity.

That is what the resistance part of

your VOM does; it provides a small electrical current (using batteries within the VOM) and measures how fast the current is flowing. For our purposes, it doesn't matter how *fast* the current is flowing; only that there *is* current flow.

To use your VOM to test continuity, set the dial on (resistance) R x 1, or whatever the lowest setting is. Touch the metal parts of the test leads together and read the meter. It should peg the meter all the way on the right side of the scale, towards "0" on the meter's "resistance" scale. If the meter does not read zero resistance, adjust the thumbwheel on the front of the VOM until it *does* read zero. If you cannot get the meter to read zero, the battery in the VOM is low; replace it.

If you are testing, say, a solenoid, first make sure that the solenoid leads are not connected to anything, especially a power source. If the solenoid's leads are still connected to something, you may get a reading through that something. If there is still live power on the item

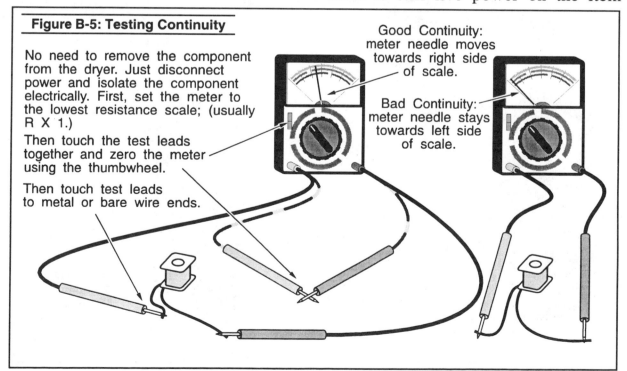

### Figure B-5: Testing Continuity

No need to remove the component from the dryer. Just disconnect power and isolate the component electrically. First, set the meter to the lowest resistance scale; (usually R X 1.)

Then touch the test leads together and zero the meter using the thumbwheel.

Then touch test leads to metal or bare wire ends.

Good Continuity: meter needle moves towards right side of scale.

Bad Continuity: meter needle stays towards left side of scale.

you're testing for continuity, you will burn out your VOM instantly and possibly shock yourself.

Touch the two test leads to the two bare wire ends or terminals of the solenoid. You can touch the ends of the wires and test leads with your hands if necessary to get better contact. The voltage that the VOM batteries put out is very low, and you will not be shocked. If there is NO continuity, the meter won't move. If there is GOOD continuity, the meter will move toward the right side of the scale and steady on a reading. This is the resistance reading and it doesn't concern us; we only care that we show good continuity. If the meter moves only very little and stays towards the left side of the scale, that's BAD continuity; the solenoid is no good.

If you are testing a switch, you will show little or no resistance (good continuity) when the switch is closed, and NO continuity when the switch is open. If you do not, the switch is bad.

## 1-4(c). AMMETERS

Ammeters are a little bit more complex to explain without going into a lot of electrical theory. If you own an ammeter, you probably already know how to use it.

If you don't, don't get one. Ammeters are expensive. And for *our* purposes, there are other ways to determine what an ammeter tests for. If you don't own one, skip this section.

For our purposes, ammeters are simply a way of testing for continuity without having to cut into the system or to disconnect power from whatever it is we're testing.

Ammeters measure the current in amps flowing through a wire. The greater the current that's flowing *through* a wire, the greater the density of the magnetic field, or *flux*, it produces *around* the wire. The ammeter simply measures the density of this flux, and thus the amount of current, flowing

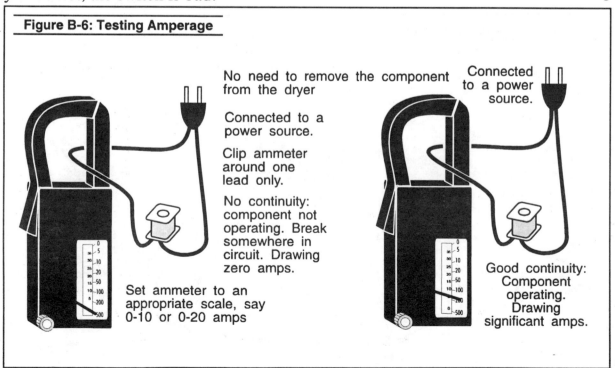

**Figure B-6: Testing Amperage**

No need to remove the component from the dryer

Connected to a power source.

Clip ammeter around one lead only.

No continuity: component not operating. Break somewhere in circuit. Drawing zero amps.

Set ammeter to an appropriate scale, say 0-10 or 0-20 amps

Connected to a power source.

Good continuity: Component operating. Drawing significant amps.

through the wire. To determine continuity, for our purposes, we can simply isolate the component that we're testing (so we do not accidentally measure the current going through any other components) and see if there's *any* current flow.

To use your ammeter, first make sure that it's on an appropriate scale (0 to 10 or 20 amps will do). Isolate a wire leading directly to the component you're testing. Put the ammeter loop around that wire and read the meter. (Figure B-6)

## 1-5. BASIC REPAIR AND SAFETY PRECAUTIONS

1) Always de-energize (pull the plug or trip the breaker on) any dryer that you're disassembling. If you need to re-energize the dryer to perform a test, make sure any bare wires or terminals are taped or insulated. Energize the unit only long enough to perform whatever test you're performing, then disconnect the power again.

I want to impress upon you something really important. In electric dryers, you're usually dealing with 220 volt circuits. DO NOT TAKE THIS LIGHTLY. I've been hit with 110 volts now and then. Anyone who works with electrical equipment has at one time or another. It's unpleasant, but unless exposure is more than a second or so, the only harm it usually does is to tick you off pretty good. However, *220 VOLTS CAN KNOCK YOU OFF YOUR FEET. IT CAN DO YOUR BODY SOME SERIOUS DAMAGE, VERY QUICKLY. DO NOT TEST LIVE 220 VOLT CIRCUITS.*

If you have a heart condition, epilepsy, or other potentially serious health conditions, well...hey, it's just my opinion, but you shouldn't be testing 220 volt circuits *at all*. It's not worth dying for.

2) If this manual advocates replacing a part, REPLACE IT!! You might find, say, a solenoid that has jammed for no apparent reason. Sometimes you can clean it out and lubricate it, and get it going again. The key words here are *apparent reason*. There is a reason that it stopped. You can bet on it. And if you get it going and re-install it, you are running a very high risk that it will stop again. If *that* happens, you will have to start repairing your dryer *all* over again. It may only act up when it is hot, or it may be bent slightly...there are a hundred different "what if's." Very few of the parts mentioned in this book will cost you over ten or twenty dollars. Replace the part.

3) If you must lay the dryer over on its side, front or back, first make sure that you are not going to break anything off, such as a gas valve. Lay an old blanket on the floor to protect the floor and the finish of the dryer.

4) Always replace the green (ground) leads when you remove an electrical component. They're there for a reason. And NEVER EVER remove the third (ground) prong in the main power plug!

5) When opening the dryer cabinet or console, remember that the sheet metal parts are have very sharp edges. Wear gloves, and be careful not to cut your hands!

6) When testing for your power supply from a wall outlet, plug in a small appliance such as a shaver or blow dryer. If you're not getting full power out of the outlet, you'll know it right away.

7) If you have diagnosed a certain part to be bad, but you cannot figure out how to remove it, sometimes it helps to get the new part and examine it for mounting holes or other clues as to how it may be mounted.

# Chapter 2

# GENERAL OPERATION AND COMPONENT DIAGNOSIS

## 2-1 DRYER BASICS & COMPONENTS

The main idea of a dryer is to circulate warm air through wet clothes to evaporate moisture from them. This sounds really simple, but there are many implications. This means that all dryers have to have a blower to move air and a heat source to warm the air, and that airflow is very important. It also means that all dryers must have a way to toss the clothes around a bit, because air won't circulate through them if they're just laying there in a big wet lump. This is done by "tumbling" them in a big round drum.

## 2-1(a) DRIVE TRAIN (Figure G-1)

All Whirlpool-designed dryers have a drive motor which turns the drum through a belt. The motor also turns a blower fan connected directly to the motor shaft.

## 2-1(b) HEAT SOURCE AND TEMPERATURE CONTROL

Warm air evaporates moisture much faster than cold air, so all dryers also have a heat source. The heat source in household machines is either gas or electric.

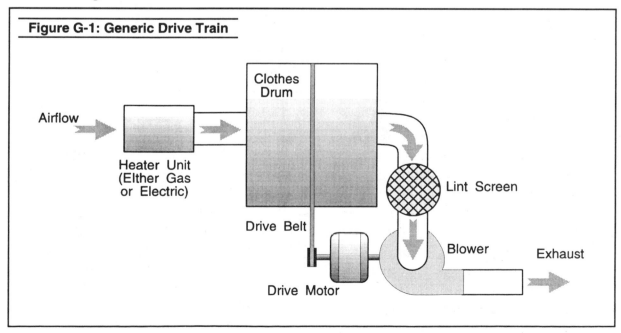

**Figure G-1: Generic Drive Train**

Airflow

Heater Unit
(Either Gas
or Electric)

Clothes
Drum

Drive Belt

Drive Motor

Lint Screen

Blower

Exhaust

Gas burners have remained remarkably similar over the years, despite going from pilot to spark to glo-bar ignition systems. Ignitor systems and holding coils on the gas safety valves have varied slightly over the years, but you can find the diagnosis and repair procedures for all gas burners covered by this manual in section 3-3.

A temperature control system keeps the air at the optimum temperature for drying and prevents scorching of your clothes. This system uses several thermostats; some adjustable and others a fixed temperature. Testing the temperature control system is discussed in section 2-2(c). Finding the thermostats is discussed in section 3-6.

There is also a safety system that prevents the heating system from starting at all unless the blower is turning. This prevents overheating of the system or natural gas buildup in the dryer cabinet. A contact in the centrifugal motor starting switch assembly only allows the heating system to heat if the motor is running.

## 2-1(c) AIRFLOW, LINT FILTER AND EXHAUST SYSTEM

Airflow is important in removing the moisture from the dryer drum; thus the blower, vents and exhaust ducting must be kept as clean (translation: free of lint) as possible. This also prevents overheating of electric heating elements and accidental lint fires, and insures enough airflow to keep the gas burner operating properly.

Also, drum seals can wear out and leak, which can disrupt proper airflow.

The lint screen traps most lint, but some does get through or leak out around the drum seals. Over a period of

years enough can build up enough to allow some of the above symptoms to occur. The airflow system is discussed in section 3-2, except for drum seals, which are discussed in section 3-4.

The blower is the last component in the airflow system. (See Figure G-1) From the blower, the air goes directly out the dryer exhaust. Thus, the other components are *not* under *pressure*. Air is being *sucked* through them, so they are under a *vacuum*.

## 2-1(d) TIMERS AND OTHER CONTROLS

Some dryer timers are simply devices that run the motor for a set number of minutes; others integrate temperature control and even moisture (humidity) controls. Yet others are solid-state, electronic units. If you have a separate timer and temperature control, consider yourself lucky; the combined units are considerably more difficult to diagnose and generally more expensive to replace. Diagnosis is discussed in Chapter 2-2(b).

## 2-2 TESTING ELECTRICAL COMPONENTS

Sometimes you need to read a wiring diagram, to make sure you are not forgetting to check something. Sometimes you just need to find out what color wire to look for to test a component. It is ESPECIALLY important in diagnosing a bad timer.

If you already know how to read a wiring diagram, you can skip this section. If you're one of those folks who's a bit timid around electricity, all I can say is read on, and don't be too nervous. It will come to you. You learned how to use a VOM in Chapter 1, right?

Each component should be labelled clearly on your diagram. Look at figure G-2. The symbols used to represent each component are pretty universal; for example, two different symbols for thermostats are shown, but both have a little square line in them, so you know they're thermostats.

A few notes about reading a wiring diagram:

Notice that in some parts of the diagram, the lines are thicker than in other parts. The wiring and switches that are shown as thick lines are *inside* of the timer.

Also note that since heaters and ignitors are both a type of resistor, they may be shown with the resistor symbol (a zigzag line) or they may have their own square symbol, but they should be clearly marked.

The small circles all over the diagram are terminals. These are places where you can disconnect the wire from the component for testing purposes.

If you see dotted or shaded lines around a group of wires, this is a switch assembly; for example, a temperature or cycle switch assembly. It may also be

the timer, but whatever it is, it should be clearly marked on the diagram. Any wiring enclosed by a shaded or dotted box is internal to a switch assembly and must be tested as described in sections 2-2(a) and 2-2(b).

Switches may be numbered or lettered. Those markings can often be found cast or stamped into the switch. To test a switch with a certain marking, mark and disconnect all the wires from your timer. Connect your ohmmeter to the two terminal leads of the switch you want to test. For example, in figure G-2, if you want to test the hi-temp selector switch, connect one lead to the M and one to the H terminal. Then flick the switch back and forth. It should close and open. If it does, you know that contact inside the switch is good.

Remember that for something to be energized, it must make a complete electrical circuit. You must be able to trace the path that the electricity will take, FROM the wall outlet back TO the wall outlet. This includes not only the component that you suspect, but all switches leading to it.

**Figure G-2: Typical Wiring Diagram**

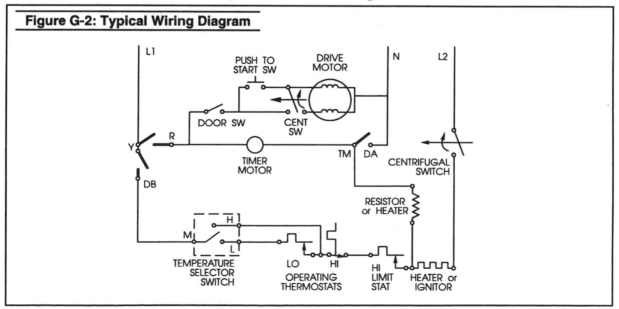

In Figure G-2(a), which shows a typical *electric* dryer, L1, L2, and N are the main power leads; they go directly to your wall plug. Between L1 and N, you will see 110 volts. Between L2 and N, you will also see 110 volts. But between L1 and L2, you will see 220 volts.

In gas dryers, L1 and L2 will be 110-volt leads. Sometimes they will be labelled L1 and N, but they are still 110 volt leads.

Let's say you need to check out why the heater is not working. Since a burnt out heater element is the most likely cause of this symptom, first test the heater for continuity. If you have good continuity, something else in the circuit that feeds the heater must be defective.

Following the gray-shaded circuit in figure G-2(a), note that the electricity flows from L1 to L2, so this is a 220-volt circuit.

From L1 the electricity flows to the Y-DB switch. This switch is located inside of the timer (you know this because it is drawn with thick lines) and it must be closed. The power then goes through the wire to the temperature selector switch. In this example, we have set the temperature on "low." Note that in this machine, on this setting, the electricity flows through both the high-temp and low-temp operating thermostats. Therefore, both thermostats must be closed and show good continuity. The electricity then flows through the high limit thermostat, so it too must be closed and show good continuity.

The electricity flows through the heater, which we have already tested and we know is good. Then the electricity flows through the centrifugal switch, which must be closed, before going back out the main power cord (L2).

To test for the break in the circuit, simply isolate each part of the system (remove the wires from the terminals) and test for continuity. For example, to test the thermostats in our example, pull the wires off each thermostat and test continuity across the thermostat terminals as described in section 2-4(c).

The Y-DB switch is shown in bold lines, so it is inside the timer. For now, let's ignore this switch. (Remember; the timer is the *last* thing you should check; see section 2-4(b).)

That leaves the centrifugal switch, which is only closed when the motor is running. However, if you can identify the proper leads, you can use your alligator jumpers to jump across them. If you do this and the heater kicks on when you turn on the timer, you know the switch is bad.

If NONE of the other components appear to be defective, test the timer as described in section 2-4(b).

To check for a wire break, you would pull each end of a wire off

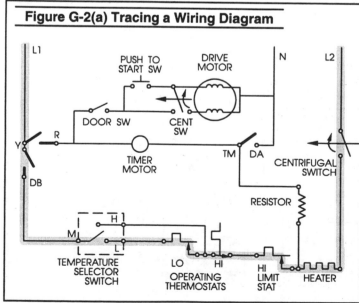

**Figure G-2(a) Tracing a Wiring Diagram**

the component and test for continuity through the wire. You may need to use jumpers to extend or even bypass the wire; for example, if one end of the wire is in the control console and the other end in underneath the machine. If there is no continuity, there is a break in the wire! It will then be up to you to figure out exactly where that break is; there is no magic way. If you have a broken wire, look along the length of the wire for pinching or chafing. If there is a place where the wires move, check there first. Even if the insulation is O.K., the wire may be broken inside.

## 2-2(a) SWITCHES AND SOLENOIDS

Testing switches and solenoids is pretty straightforward. Take all wires off the component and test resistance across it.

Switches should show good continuity when closed and no continuity when open.

Solenoids should show SOME resistance, but continuity should be good. If a solenoid shows no continuity, there's a break somewhere in the windings. If it shows no resistance, it's shorted.

## 2-2(b) TIMERS

The timer is the brain of the dryer. It controls everything in the cycle. In addition to telling the motor when to run, it may also activate the heating circuit or heating control circuits, humidity-sensing circuits, etc.

Solid state timers are difficult and expensive to diagnose. If you suspect a timer problem in a solid-state system, you can try replacing it, but remember that it's expensive and non-returnable (being an electrical part.) If you have

one of these units that's defective, you can check into the cost of replacing it, but it's been my experience that you usually will end up just replacing the whole dryer or calling a technician. If you do call a technician, make sure you ask up front whether they work on solid-state controls.

Most timers are nothing more than a motor that drives a set of cams which open and close switches. Yet it is one of the most expensive parts in your dryer, so don't be too quick to diagnose it as the problem. Usually the FIRST thing a layman looks at is the timer; it should be the LAST. And don't forget that timers are electrical parts, which are usually non-returnable. If you buy one, and it turns out *not* to be the problem, you've just wasted the money.

In a wiring diagram, the wiring and switches that are inside the timer will usually be drawn with dark lines.

## TIMER DIAGNOSIS

If the timer is not advancing *only in the automatic or humidity control cycle*, (i.e. more dry-less dry on your timer dial) see section 2-2(c).

If the timer is not advancing in all cycles, well, that's pretty obvious. Replace the timer or timer drive motor, or have it rebuilt as described below.

Timers can be difficult to diagnose. The easiest way is to go through everything else in the malfunctioning system. If none of the other components are bad, then it *may* be the timer.

Remember that a timer is simply a set of on-off switches. The switches are turned off and on by a cam, which is driven by the timer motor. Timer wires are color-coded or number-coded.

Let's say you've got a motor starting problem. Following the shaded circuit in figure G-3, you test the door switch, push-to-start switch and centrifugal switch. They all test ok. So you think you've traced the problem to your timer.

First unplug the machine. Looking at your wiring diagram, you see that the Y-R switch feeds the drive motor. RE-MOVE those wires from the timer and touch the test leads to those terminals. Make sure the timer is in the "on" position and slowly turn the timer all the way through a full cycle. (On some timers, you cannot turn the dial while it is on. You must simply test the timer one click at a time. Be patient!)

You should see continuity make and break at least once in the cycle; usually several times. If it doesn't, the internal contacts are bad; replace the timer.

In general, timers cannot be rebuilt by the novice. Check with your parts dealer; if it *can* be rebuilt, he'll get it done for you. If it's a common one, your parts dealer may even have a rebuilt one in stock.

For the most part, if your timer is acting up, you need to replace it. To replace, mark the wires or note the color codes written on the timer. If you need to, you can draw a picture of the terminal arrangement and wire colors. If possible, change over the timer wires one-by-one. It can be easier. If there are any special wiring changes, they will be explained in instructions that come with the new timer.

## 2-2(c) THERMOSTATS AND OTHER TEMPERATURE CONTROLS, AND HUMIDITY SENSORS

### THERMOSTATS
(Figure G-4)

A thermostat is basically just a switch that opens or closes according to the temperature that it senses. This switch can be used to turn the heating system (either gas or electric) on and off to maintain a certain air temperature range at a given place in the system.

You can test it as described in section 1-4(b) by testing for continuity across its terminals. A cold *operating* stat or *hi-limit* stat should show continuity. A cold cool-down stat should show no continuity. There may be several different thermostats side by side; for example 135 degrees for low temperature, 165 for high temperature, etc. You choose which thermostat is used by selecting the heat setting on the dryer console.

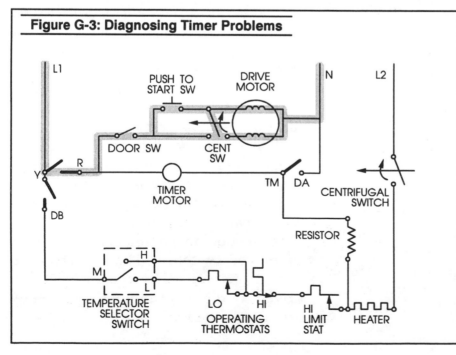

**Figure G-3: Diagnosing Timer Problems**

## AUTO CYCLES (More Dry-Less Dry)

In an "auto" cycle, the system has some way of sensing how much moisture is in the air inside the dryer drum. If the air is dry, the timer advances more quickly to end the cycle sooner. This is done in one of two ways.

When the air in the drum is moist, the water in it absorbs heat to evaporate. This keeps the air temperature lower, and it takes longer to heat up. The thermostat on the drum exhaust will keep the heating system on longer. In these systems, that same thermostat controls the timer motor; while the heating system is on, the timer motor is not running, and vice-versa. As the clothes get drier, the exhaust air temperature increases more quickly and stays hotter, so the heating system doesn't stay on for as long, and the timer motor runs more, and the cycle ends sooner.

The thermostats in these machines have three leads. One side goes to the timer motor, the other to the heating system. These thermostats are also difficult to test without any way to heat them up. But they're pretty cheap. If the symptoms lead you to suspect that yours is defective, just replace it.

Besides heating more slowly, moist air also conducts electricity better than dry air. So another way the engineers design a humidity sensor is to put two electrical contacts inside the dryer drum. (see figure W-9 in chapter 3) The electrical currents conducted by the air are so low that an electronic circuit is needed to sense when the air is moist, but essentially the same thing happens in this system as in the other. When the air is moist, the timer motor doesn't run as often. When the air is dry, the timer motor runs longer and times out sooner.

The sensors in these machines tend to get coated with gummy stuff, especially if you use a lot of fabric softener in the wash or starch in ironing. If the timer is not advancing during the auto cycle, this is likely what has happened. Try scrubbing the sensors with a little 409 to get it off. In extreme cases, use a little rubbing alcohol as a solvent. The circuit board could have gone bad, too; there is no good way to test it without a lot of expensive equipment. If you think it has gone bad, it probably has. Replace it.

**SPECIAL NOTE:** *In electric dryers with an automatic cycle, a special problem exists. The problem is that the heater operates on 220 volts, but the timer motor runs on 110 volts. There is a resistor in the system to cut down the voltage (see figure G-3 and if this resistor is bad, you will see the same symptoms as if the thermostat was bad: the timer motor will not run in the automatic cycle. If you have one of these dryers, make sure you test the resistor for continuity, in addition to the thermostat.*

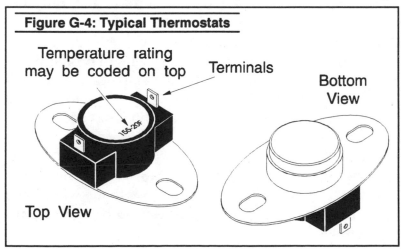

**Figure G-4: Typical Thermostats**

Temperature rating may be coded on top

Terminals

Bottom View

155-20F

Top View

If a thermostat fails into a closed position, there is a danger that the heating system will continue operating until something catches fire. To prevent this, there is a high-temperature thermostat that will cut out the entire heating system.

There is no way to repair thermostats. Replace any that are bad.

## OTHER TEMPERATURE CONTROLS

Selecting which thermostat is used may be done inside the timer, or there may be a separate multi-switch that accomplishes this. Test internal timer switches as described in section 2-2(b).

Separate temperature selector switches are tested either by measuring continuity through each contact (figure G-5) or by jumping across the two correct terminals with your alligator jumpers.

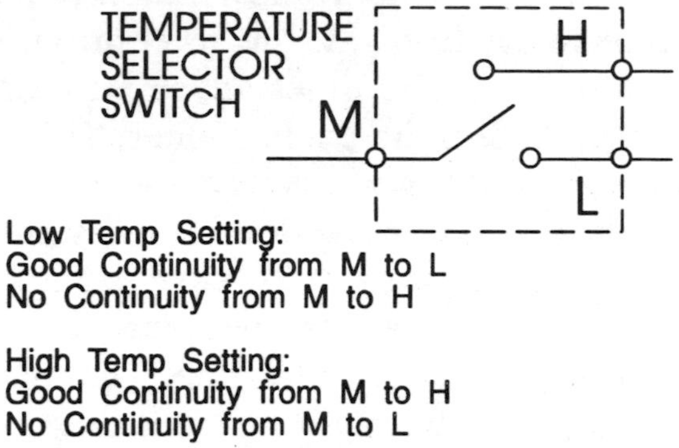

## 2-2(d) DRIVE MOTORS AND CENTRIFUGAL START SWITCH

A motor that is trying to start, but can't for whatever reason, is using one heckuva lot of electricity. So much, in fact, that if it is allowed to continue being energized in a stalled state, it will start burning wires. To prevent this, an overload switch is installed on motors to cut power to them if they don't start within a certain amount of time.

If the motor is trying to start, but can't, you will hear certain things. First will be a click, followed immediately by a buzzing sound. Then, after about 5 to 20 seconds of buzzing, another click and the buzzing will stop. The sounds will keep repeating every minute or two. In some extreme cases, you may even smell burning.

If you hear the motor doing this, but it won't start, disconnect power and take

**Figure G-5: Testing a Typical Selector Switch**

TEMPERATURE SELECTOR SWITCH

Low Temp Setting:
Good Continuity from M to L
No Continuity from M to H

High Temp Setting:
Good Continuity from M to H
No Continuity from M to L

all the load off it. For example, disconnect the drive belt, and make sure nothing is jamming the blower wheel. The motor should turn easily by hand.

Try to start the motor again. If it still won't start, the motor is bad. If you have an ammeter, the stalled motor will be drawing 10 to 20 amps or more.

## STARTING SWITCH

Dryers have a centrifugal starting switch mounted piggyback on the motor. There are many sets of contacts inside the switch, and each design is different, even among dryers of the same brand. Testing the switch is most easily accomplished by replacing it.

Remember that starting switches are electrical parts, which are generally not returnable. If you test the switch by replacing it, and the problem turns out to be the motor itself, you will probably not be able to return the starting switch for a refund. But they're pretty cheap, and if it *is* the problem, you just saved yourself the best part of a hundred bucks for a new motor.

If the motor is stalled (buzzing and/or tripping out on the overload switch) and the starting switch tests O.K., the motor is bad. Replace it.

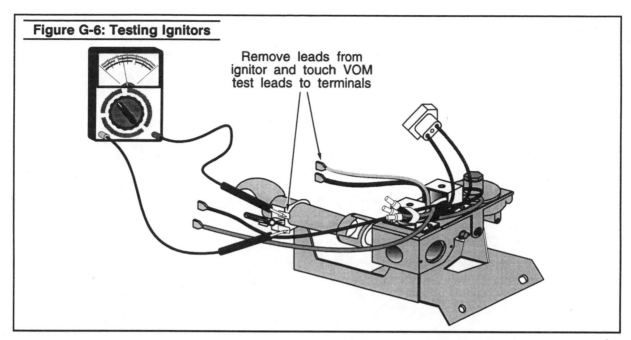

**Figure G-6: Testing Ignitors**

Remove leads from ignitor and touch VOM test leads to terminals

*NOTE: A replacement motor may come without a pulley. When buying a new motor, make sure that the pulley can be changed over, or else get a new pulley with the new motor. It may save you a second trip to the parts dealer.*

## 2-2(e) IGNITORS

You can test the ignitor by testing for resistance across the element. (See figure G-6) A good ignitor will show quite a bit of resistance, between about 50 and 600 ohms. A bad ignitor will usually show no continuity at all.

## 2-2(f) ELECTRIC HEATERS

Electric heater elements are tested by measuring continuity across them. Like ignitors, they should show quite a bit of resistance, and defective heaters will usually show no continuity at all.

## 2-3 DRIVE BELTS AND TENSIONERS

To access the drive belt, see chapter 3.

If you see any of the problems pictured in figure G-7, replace the belt.

It is important for the belt tensioner to be operating properly. Check for broken springs. Also check that the tensioner idler roller spins freely.

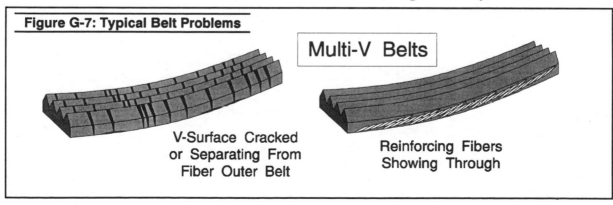

**Figure G-7: Typical Belt Problems**

Multi-V Belts

V-Surface Cracked or Separating From Fiber Outer Belt

Reinforcing Fibers Showing Through

# Chapter 3

# *DIAGNOSIS & REPAIR*

## 3-1 MACHINE LAYOUT

Figure W-1 shows the general layout of these machines.

The drum is supported at the front by a slider around the inside lip of the drum, and at the rear by support rollers.

The blower fan is attached directly to the rear of the drive motor. The drive belt comes directly off the front of the motor and goes completely around the drum.

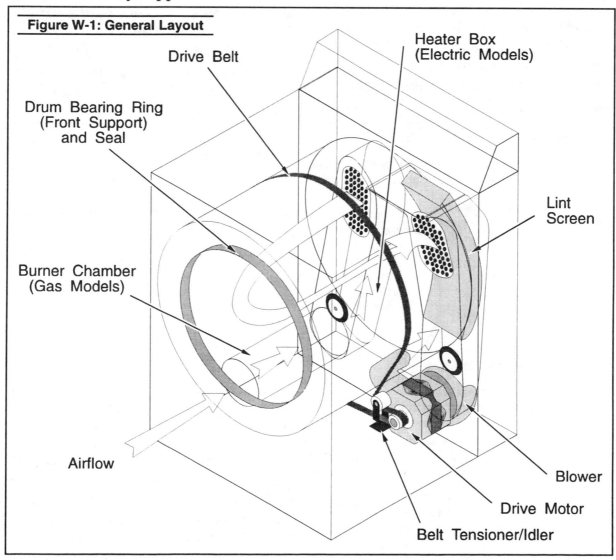

**Figure W-1: General Layout**

Drive Belt

Heater Box (Electric Models)

Drum Bearing Ring (Front Support) and Seal

Lint Screen

Burner Chamber (Gas Models)

Airflow

Blower

Drive Motor

Belt Tensioner/Idler

## 3-2 COMMON PROBLEMS & DIAGNOSES

The most common problems in these machines are:

### NOISY OPERATION

Usually a loud rumbling sound getting progressively worse over time, caused by worn drum support rollers. Usually this appears in machines about 7-15 years old. To replace the rollers, remove the drum as described in section 3-4.

Less likely, but still possible, you may also get a rumbling noise if a belt tensioner pulley seizes up. Remove belt tension as described in section 3-4 and check that the idler pulley spins freely. If not, replace the tensioner.

If you hear a loud, regular clackety-clacking as the dryer drum is turning, and you do not have any metal zippers or buttons inside the drum, some coins may have gotten inside the vanes. Look inside the dryer drum. One of the three tumbling vanes will be plastic. Turn the drum until the plastic vane is on top. Open the top of the cabinet as shown in figure W-2. Remove the screws that hold the vane in place and remove any coins that have gotten in the vanes.

### BROKEN DOOR CABLES

Another fairly common problem is that the dryer door support cables break. See figure W-3 for a cross section of the cable mechanism. To access the cables, lift the top of the cabinet as shown in figure W-2 and remove the kickplate. In models without a kickplate, you will need to remove the whole front of the cabinet.

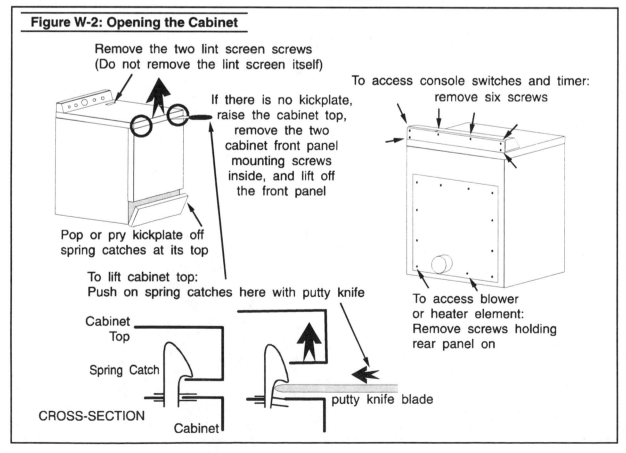

**Figure W-2: Opening the Cabinet**

Remove the two lint screen screws
(Do not remove the lint screen itself)

If there is no kickplate, raise the cabinet top, remove the two cabinet front panel mounting screws inside, and lift off the front panel

To access console switches and timer: remove six screws

Pop or pry kickplate off spring catches at its top

To lift cabinet top:
Push on spring catches here with putty knife

To access blower or heater element: Remove screws holding rear panel on

Cabinet Top

Spring Catch

CROSS-SECTION

Cabinet

putty knife blade

## NO HEAT

No heat, caused by a broken ignitor or other heating system problems, or vent problems. Diagnose and replace as described in section 3-3 & 3-5, and check thermostats as described in section 2-2(c).

## DRYER DRUM IS NOT TURNING

If the drum is not turning, there may be no noise or other external symptoms. The clothes will simply be laying there in a big wet lump and they won't dry. The dryer probably won't *sound* normal either. To diagnose, start the machine empty, open the door and look inside quickly, or depress the door switch to see if the drum is turning. If not, the belt or belt tensioner is usually broken. To replace, see section 3-4.

You may see similar symptoms if the motor has gone bad, except that you probably will not hear the motor turning. If the motor is locked, you may hear it buzzing. See section 2-2(d) about motors. If the motor won't run it may also be the thermal fuse; see section 3-6.

## DRYER ISN'T HEATING, OR CLOTHES AREN'T DRYING, OR CLOTHES ARE TAKING TOO LONG TO DRY

This *can* be caused by poor airflow in all dryers, but *especially* in gas dryers. Feel the dryer vent exhaust (usually outside the house.) If there isn't a strong blast of air coming out, check the lint screen and open up any dryer vent you can get to to check for clogging. Also check any flexible dryer vent

for pinching. Remove the vent from the back of the dryer and feel the air blast without the exhaust system connected. If there is a big difference in how the airstream feels, you probably have a vent problem.

If your drum is turning and you have good airflow, see section 3-3 about gas burners or section 3-5 about electric heater elements.

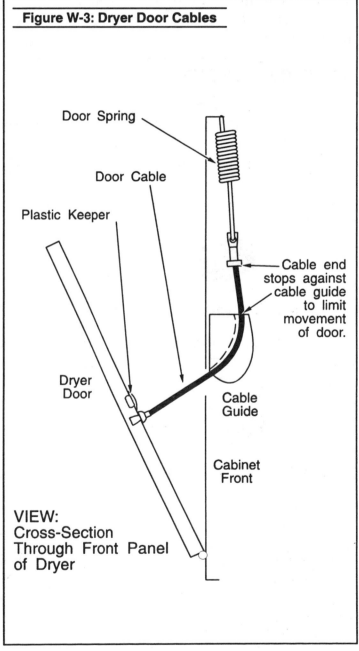

**Figure W-3: Dryer Door Cables**

Door Spring

Door Cable

Plastic Keeper

Cable end stops against cable guide to limit movement of door.

Dryer Door

Cable Guide

Cabinet Front

VIEW: Cross-Section Through Front Panel of Dryer

## 3-3 GAS BURNERS

To access the burner assembly, remove the front bottom panel (kickplate) of the dryer by tugging sharply at the top of it. (Figure W-2)

Gas burners can generally be divided into two broad categories; pilot and pilotless ignition. Pilot ignition models have not been manufactured in a number of years, and thus tend to be older units, but there are still a significant number of them in operation.

The main components of the system (see Figure W-4) are the gas valve, venturi and burner chamber, and gas safety solenoids (in all models,) the flame sensor and the ignitor (in pilotless models,) or the pilot orifice and sensor (in pilot models.)

### IGNITORS

See figure W-5 for pictures of some different types of ignitors.

Most ignitors just glow until the mechanism opens the gas valve. The

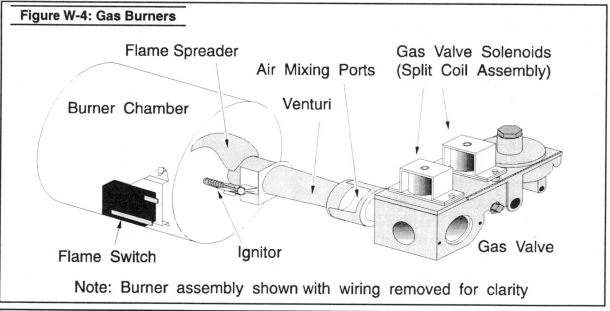

**Figure W-4: Gas Burners**

Flame Spreader

Air Mixing Ports

Gas Valve Solenoids (Split Coil Assembly)

Burner Chamber

Venturi

Flame Switch

Ignitor

Gas Valve

Note: Burner assembly shown with wiring removed for clarity

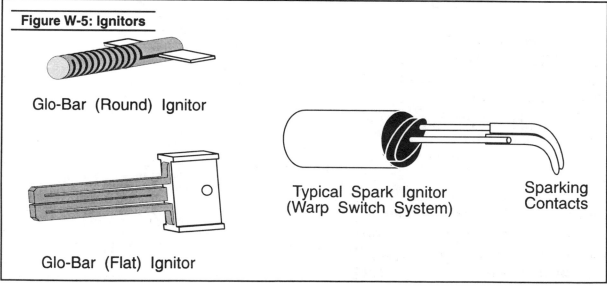

**Figure W-5: Ignitors**

Glo-Bar (Round) Ignitor

Glo-Bar (Flat) Ignitor

Typical Spark Ignitor (Warp Switch System)

Sparking Contacts

"warp switch" ignitor sparks to ignite the gas (see "NORMAL BURNER OPERATION" below.) If you have a "glo-sil" ignitor, (it looks like a tiny coil) and it is burnt out, you are probably out of luck. They are difficult and expensive to get parts for, and considering the age of these dryers, you will probably end up getting a new dryer.

## NORMAL BURNER OPERATION

A normally operating gas burner system will have a clean, mostly blue flame (perhaps occasionally streaked with just a little tinge of orange) that cycles on and off every couple of minutes.

When the flame is off and starts to cycle on, you will hear a loud click. In a pilot system, the gas valve will open and the flame will kick on at this time. In a pilotless system, the ignitor will heat up and glow brightly for about 7-15 seconds. The flame sensor senses the heat from the ignitor. If it is glowing, you will hear another click, the gas valve will open and the flame will kick on. This is a safety feature; if you don't have ignition, you certainly don't want to open the gas valve and dump gas into the dryer cabinet.

Models with a spark ignitor, or "warp switch" ignitor act slightly differently. Instead of the ignitor glowing, you will hear a metallic rattling sound and see what looks almost like a welder's arc flashing. The flame starts immediately and after about 15-20 seconds, the metallic rattling will stop but the flame will stay on.

When the air in the dryer drum reaches the chosen temperature (as sensed by the thermostats) the gas valve will close and the flame will shut off.

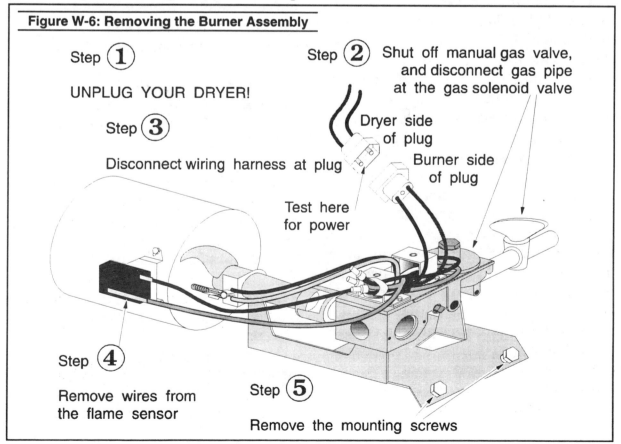

**Figure W-6: Removing the Burner Assembly**

Step ① UNPLUG YOUR DRYER!

Step ② Shut off manual gas valve, and disconnect gas pipe at the gas solenoid valve

Step ③ Disconnect wiring harness at plug

Dryer side of plug

Burner side of plug

Test here for power

Step ④ Remove wires from the flame sensor

Step ⑤ Remove the mounting screws

## 3-3(a) GAS BURNER SYMPTOMS AND REPAIR

### IGNITOR DOESN'T GLOW (PILOTLESS SYSTEMS ONLY)

Nine times out of ten, if the ignitor is not glowing, it is burnt out. Unplug the dryer and remove the burner assembly as shown in figure W-6. Usually you will see a white or yellowish burned area and a break in the ignitor. If so, replace the ignitor. If you can't see an obvious break, test the ignitor for continuity as shown in section 2-2(e).

If the ignitor is not defective, you need to isolate whether the problem is in the control (thermostat) area, or at the burner itself. The general idea is that if you have 110 volts getting to the burner assembly, then the thermostats are OK, and something in the burner assembly is bad. If you don't have 110 volts at the burner, then a thermostat or some other control is bad.

Perform the following test:

1) Unplug the dryer and open the burner inspection cover.

2) Unplug the main wiring harness leading to the burner assembly. (Figure W-6)

3) Using your alligator jumpers, connect a VOM to the *dryer* side of the harness plug. (as opposed to the side of the harness plug that's connected to the *burner*.)

4) Make sure no wires will get caught up in the turning drum. Set your dryer timer to the "on" position, high heat, and plug in the dryer. If your VOM reads 110 volts, something in the burner assembly is bad. If the burner *isn't* getting voltage, the problem is in one of the components of the heating control system: a thermostat, timer or temperature control switch, or motor centrifugal switch.

If you trace it to the burner assembly, and you've already eliminated the ignitor as the problem, either the flame sensor or the gas valve solenoid coil(s) are bad. Unplug the dryer, disconnect the flame sensor and test it for continuity. If you have no continuity, the flame sensor is bad. If you have continuity, the coils are bad. Bring the burner assembly to your parts dealer to make sure you get the right coil assembly, and don't forget to bring the model number of the machine.

### SPARKER DOESN'T SPARK, OR FLAME DOESN'T START (WARP SWITCH SYSTEM ONLY)

In the older models mentioned previously, if you do not hear the metallic rattling sound, the ignitor may be out of adjustment. It can be a difficult and complex adjustment to make. Take the burner assembly to your parts dealer and they should be able to help you adjust it or guide you to someone who can.

### IGNITOR GLOWS, BUT FLAME DOES NOT START (GAS VALVE DOES NOT OPEN) (PILOTLESS SYSTEMS ONLY)

Either the flame sensor is not working properly or the safety solenoid coils are not opening the gas valve. If the ignitor doesn't stop glowing, the flame sensor is bad. If the ignitor cycles on and off, the gas solenoid coil(s) are bad.

### FLAME STARTS (GAS VALVE OPENS) BUT KICKS OFF QUICKLY (SHORT-CYCLES, LOW HEAT)

Usually this problem can be traced to airflow problems, especially if the flame is very orangey-colored while it is on (rather than blue.) The solution is to clean out your lint screen or dryer exhaust. It is an especially common problem in installations where the dryer exhaust runs a long way before venting to the outside. Check as described in section 3-2.

This may also be caused by a defective flame sensor. Test as described in the section above, "IGNITOR DOESN'T GLOW".

Occasionally this problem can be caused by a bad thermostat. Test as described in section 2-2(c).

## PILOT SYSTEM PROBLEMS
### *PILOT WON'T STAY LIT (PILOT SYSTEM ONLY)*

If the pilot won't stay lit, the pilot & unlatch assembly is usually defective.

The pilot & unlatch assembly (figure W-7) is a safety mechanism. If there is no pilot, it closes off the gas to both the pilot and the main gas valve. This prevents accidental buildup of gas in the dryer cabinet.

The sensor is simply a bulb, like a thermometer bulb, with a liquid inside that expands when the pilot flame heats it. The liquid pushes against a metal diaphragm that holds the spring-loaded gas valve open. If the pilot goes out, the liquid cools and the diaphragm lets the spring close the gas valve.

In order to light the pilot, you must manually hold the valve open for a minute or so, until the sensor heats up enough to hold the valve open.

### *PILOT BURNS, BUT FLAME DOES NOT START (PILOT SYSTEM ONLY)*

The gas valve solenoid coil is not opening the main gas valve. Either the gas valve solenoid coil is defective, or the burner is not getting the signal to start burning from the heating control system.

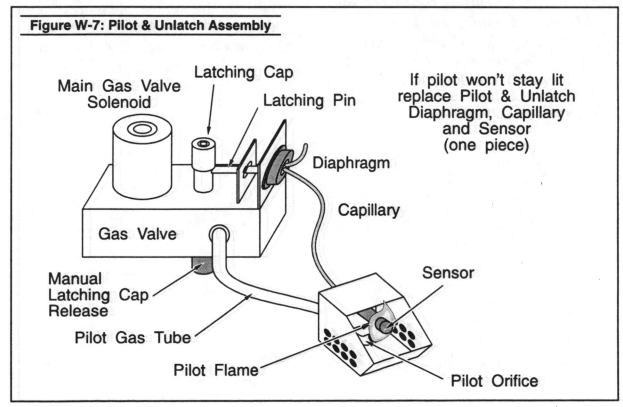

**Figure W-7: Pilot & Unlatch Assembly**

Main Gas Valve Solenoid
Latching Cap
Latching Pin
Diaphragm
Capillary
Gas Valve
Manual Latching Cap Release
Pilot Gas Tube
Pilot Flame
Sensor
Pilot Orifice

If pilot won't stay lit replace Pilot & Unlatch Diaphragm, Capillary and Sensor (one piece)

Test the wiring as described previously in the "IGNITOR DOESN'T GLOW" section to see if it is getting the 110-volt signal from the heating control system.

If so, the gas solenoid valve is bad. Replace it.

If the burner isn't getting 110 volts, something is wrong with the heating control system as described previously in the "IGNITOR DOESN'T GLOW" section.

## 3-4 DRUM REMOVAL

Unplug the machine and move it away from the wall far enough so the console will not hit the wall when you raise the cabinet top.

Lift the lint screen access cover and remove the two screws inside.

*NOTE: Do not remove the lint screen unless necessary, and only after removing the screws. This will prevent your accidentally dropping the screws down the lint screen slot. If you accidentally drop anything down the lint screen slot, it goes right into the blower wheel. If*

*this happens, see section 3-6 for access to the blower wheel.*

Lift the top of the cabinet as shown in figure W-2. If you have a model with a kickplate, remove the kickplate, too.

If you have a kickplate model, remove the belt from the motor pinion. The easiest way is to push the belt tensioner to loosen the belt with your thumb, and use the fingers of the same hand to slip the loop of belt off the motor pinion. (See figure W-8)

Loosen (but don't remove) the bottom screws holding the front of the cabinet on. Remove the two top screws inside the cabinet that hold on the cabinet front.

*CAUTION: Remember that the front of the cabinet supports the front of the dryer drum! Remove the cabinet front carefully!*

Balance the whole shebang against your knees and disconnect the wire leads from the door switch. Hold the dryer drum in place and lift the front of the cabinet off.

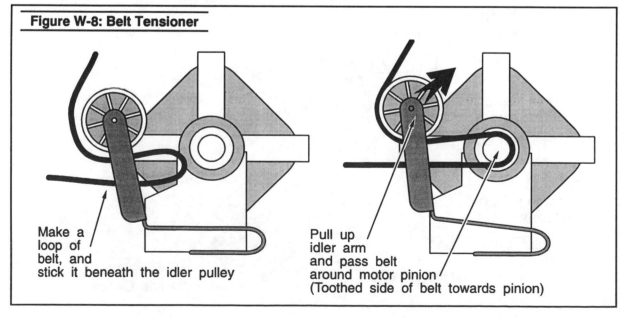

**Figure W-8: Belt Tensioner**

Make a loop of belt, and stick it beneath the idler pulley

Pull up idler arm and pass belt around motor pinion (Toothed side of belt towards pinion)

*NOTE: In certain drop-door models, (known in the parts houses as "hamper-door" models) there may be a couple of extra screws holding on the front of the cabinet, in the middle of the hinges.*

To remove the dryer drum from the cabinet:

1) In kickplate models, you have already removed the belt tension, so slide the drum straight outward through the curved cutouts in the cabinet sides. (See figure W-9)

2) In models without a kickplate, you must hold the dryer drum in place while you reach beneath it to remove the belt

tension as described previously. It takes a little acrobatics, but it isn't *too* tough.

The drum support rollers will now be easily accessible. (See figure W-9) The one on the left tends to be worn more than the other one, but replace them both. When replacing, use only ONE drop of oil on the hub. Oil tends to attract dust and lint, and over-oiling them can actually shorten the life of them.

While you have the dryer dismantled, vacuum out all the dust you can. Also, it's a good idea to replace the belt and belt tensioner whenever you have the dryer dismantled to this point. It is cheap insurance against future problems.

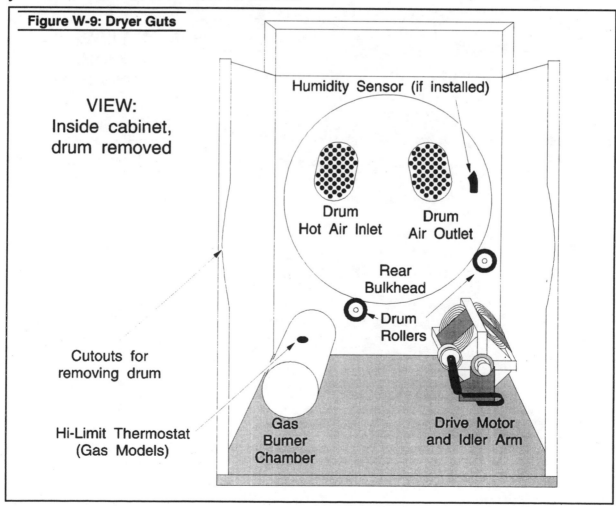

**Figure W-9: Dryer Guts**

VIEW: Inside cabinet, drum removed

Humidity Sensor (if installed)

Drum Hot Air Inlet

Drum Air Outlet

Rear Bulkhead

Drum Rollers

Cutouts for removing drum

Hi-Limit Thermostat (Gas Models)

Gas Burner Chamber

Drive Motor and Idler Arm

**HINT:** *Coins that find their way out the rear drum seal tend to end up directly beneath the left drum roller. On one job, I found eight bucks worth of quarters in one machine! Two kids, video arcade game freaks...you know the story...and YES, I showed them to mom, and she got to keep them. Hey, I may not be too smart, but I'm honest!*

Inspect the rear drum seal (attached to the back of the dryer drum.) If it is badly worn, replace it. Your parts dealer will have a seal kit, and it is actually quite inexpensive and easy to glue a new one in place.

Re-assembly is basically the opposite of disassembly.

Installing the drum can be a bit tricky, especially if you don't have a kickplate machine. Make sure the belt is around the drum before you put the drum in place. Before you put the cabinet front on, reach beneath the drum and make sure the two rollers are in the groove of the drum. If you have a kickplate machine, you can install the cabinet front at this time. If you don't have a machine with a kickplate, you have to line up and put tension on the belt *before* you put on the cabinet front.

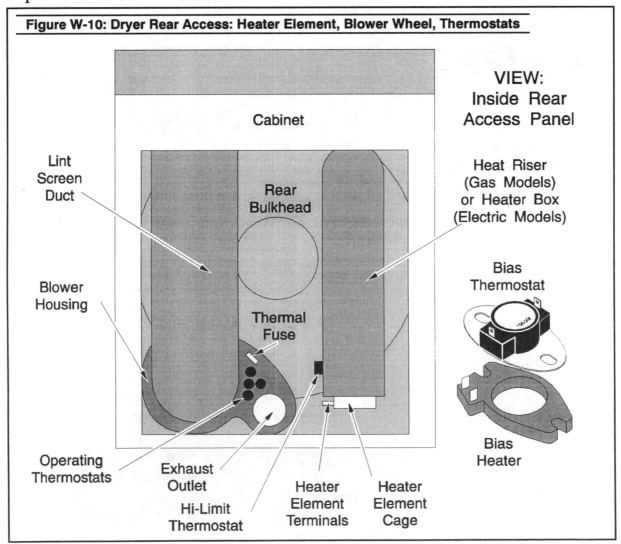

**Figure W-10: Dryer Rear Access: Heater Element, Blower Wheel, Thermostats**

When the drum is in place, make sure the belt is flat on the drum and closely line up the belt on on the old belt skid mark on the drum. To tension the belt, make a loop of belt and stick it under the tensioner as shown in figure W-8. Push the tensioner with your thumb and slip the belt loop around the motor pinion.

After installing the dryer front, roll the dryer drum around several times by hand. As you turn it, check that the belt is flat all the way around. Also check that the rear drum seal is flat all the way around, and not pinched between the drum and the rear support plate.

### 3-5 REPLACING THE HEATER ELEMENT (ELECTRIC MODELS)

The heater element is accessed through the back of the dryer, as shown in figure W-10.

The heater elements are located inside the riser on the right, as you look at the back of the machine. You can test it for continuity without removing it. To remove the riser, raise the top of the cabinet and disconnect the strap that holds the top of the riser.

### 3-6 ACCESS TO THE BLOWER WHEEL, THERMOSTATS, AND THERMAL FUSE

To get to the blower wheel, remove the back of the dryer as shown in figure W-10.

The blower wheel is under the bottom of the lint screen duct to your left, as you look at the back of the dryer. To get to the blower wheel, you must of course remove the screws holding the duct to the blower housing, but you also must remove the two screws holding the duct to the top of the cabinet, and lift the cabinet top. These two screws are inside the lint screen door. I recommend leaving the lint screen in place; if you accidentally drop one of the screws, you may have to pull the blower wheel to get it out, where you may not have had to otherwise.

The operating thermostats and thermal fuse are at the blower wheel outlet.

In most models, there are two or three operating thermostats at the blower outlet. In some, there is one thermostat nestled into a bias heater. The bias heater is a tiny, 3 to 4 watt heater that heats the thermostat to make it cut the heating system out at a lower temperature. If the dryer seems to be operating too hot in certain cycles, the bias heater may be bad. Test for resistance; it should be about 3200 to 5300 ohms.

The thermal fuse is a secondary protection in case one of the thermostats doesn't work. It is in series with the drive motor, so if the motor won't run it may be the thermal fuse. If the thermal fuse blows, replace it, along with the operating thermostats and hi-limit stat.

In electric models, the hi-temperature cutout thermostat is mounted to the heat riser; in gas models, on top of the burner chamber.

# *Index*

# Also Available from EB Publishing
## Brand-Specific Dryer & Top-Loading Washer Manuals!
*for those who want a little LESS...*
*Our brand-specific manuals have the same high quality instructions and illustrations*
*as our all-brand manuals, at a new low price!*

**2000 Edition**
### Whirlpool Dryer Repair
ISBN 1-890386-42-1        Part No. EBWD

**2000 Edition**
### Whirlpool Washer Repair
ISBN 1-890386-41-3        Part No. EBWW

*Whirlpool-brand manuals include*
*Kenmore, Kitchenaid, Estate*
*and Roper Brand Machines*

**2000 Edition**
### GE/Hotpoint Dryer Repair
ISBN 1-890386-44-8          Part No. EBGD

**2000 Edition**
### GE/Hotpoint Washer Repair
ISBN 1-890386-43-X          Part No. EBGW

*GE-brand manuals include*
*Hotpoint, late-model RCA, and*
*JC Penney (Penncrest) Brand Machines*

**2000 Edition**
### Maytag Dryer Repair
ISBN 1-890386-46-4      Part No. EBMD

**2000 Edition**
### Maytag Washer Repair
ISBN 1-890386-45-6      Part No. EBMW

# *Available from EB Publishing*

### *Cheap and Easy!* CLOTHES DRYER REPAIR
Part No. EBHD
ISBN 1-890386-03-0

Written especially for folks who are a bit mechanically-minded, but just need a little coaching when it comes to appliances. Our dryer manual covers the most common problems with the most common machines, including Whirlpool / Kenmore, GE, Maytag, Hotpoint, JC Penney, Amana, Norge, Speed Queen, Westinghouse, Kelvinator, Gibson, Frigidaire, Montgomery Ward, Signature and many others. Diagnosis procedures for gas burner and electric heater problems are covered in detail, as well as electrical, broken belts, restricted airflow and other common breakdowns.

### *Cheap and Easy!* WASHING MACHINE REPAIR
Part No. EBWM
ISBN 1-890386-02-2

Did you know that if a Maytag washer is leaking, 90% of the time, it's coming from one certain part, which costs about $3 to replace? Ten percent of all washer parts account for ninety per-cent of all washer repairs. But which ten? In this best-seller, we tell you! Why waste your time reading about rebuilding transmissions if, as a novice, you'll never do it? Why not go straight to the problem, fix it, and get back to what you were doing? Our washer manual covers the most common problems (such as leaks and other water problems, electrical and timer problems, belt and drive train problems, etc.) with the most common machines, including Whirlpool / Kenmore (both belt drive and direct drive,) GE, Maytag, Hotpoint, JC Penney, Amana, Norge, Speed Queen, Kelvinator, Frigidaire, Westinghouse, Gibson, Montgomery Ward, Signature and many others.

### *Cheap and Easy!* REFRIGERATOR REPAIR
Part No. EBHR
ISBN 1-890386-01-4

Tired of refrigerator manuals that talk about thermodynamic theories, ad nauseam? Don't you really just want to know basically how the bloomin' thing works, then get on with fixing it? Written especially for mechanical-minded folks who just need a little coaching when it comes to refrigera-tors, our repair manual cuts out all the unnecessary theory and sealed system repairs that the novice will never perform. Instead, it focuses on diagnosis and repair of the most common problems (less than 5% of all repairs have anything to do with the sealed system, or Freon.) Most refrigera-tor repairs can be accomplished in one afternoon, with no special tools or knowledge other than this manual. Includes Whirlpool / Kenmore, GE, Hotpoint, JC Penney, Amana, Norge, Westinghouse, Philco, Frigidaire, Gibson, Kelvinator, Montgomery Ward, Signature and many othe

### *Cheap and Easy!* DISHWASHER REPAIR
Part No. EBDW
ISBN 1-890386-04-9

Why are some dishwashers so bloomin' sensitive to food left on the dishes, when it seems like others could double as a wood chipper? Why do some dishwashers leave dishes so clean, when others leave spots as heavy as a new snowfall? Our dishwasher manual ponders such profound questions, as well as more acute ones, like: Where the heck is that water coming from? Written especially for do-it-yourselfers that just need a little coaching when it comes to appliances, our dishwasher manual covers the most common problems with the most common machines, including Whirlpool / Kenmore, GE, Maytag, Hotpoint, JC Penney, Frigidaire, Westinghouse, and a host of others.

### *Cheap and Easy!* OVEN & COOKTOP REPAIR
Part No. EBOC
ISBN 1-890386-05-7

Don't know the difference between a thermostat valve and a safety valve? We'll help you figure it out! Covers both gas and electric models of ovens and cooktops, ranges and stoves, self-cleaners and convection ovens. Covers safety valves and thermostatic controls, timers and other automatic controls. Includes diagnosis and repair procedures for electrical problems both 110 and 220 volt, burner and pilot flame troubles, pilot ignition including millivolt system and two-level pilots, and pilotless ignition systems including spark, re-ignition and glow-bar ignition.